Student's Solutions Manual

to accompany

Earth Algebra
Second Edition

Student's Solutions Manual

to accompany

Earth Algebra
Second Edition

Schaufele • Zumoff • Sims • Sims

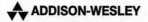

An imprint of Addison Wesley Longman, Inc.

Reading, Massachusetts • Menlo Park, California • New York • Harlow, England
Don Mills, Ontario • Sydney • Mexico City • Madrid • Amsterdam

Reproduced by Addison Wesley Longman from camera copy supplied by the authors.

Copyright © 1999 Addison Wesley Longman.

All rights reserved. No part of this publication may be reproduced, stored in a retrieval system, or transmitted, in any form or by any means, electronic, mechanical, photocopying, recording, or otherwise, without the prior written permission of the publisher. Printed in the United States of America.

ISBN 0-321-02856-2

1 2 3 4 5 6 7 8 9 10 VG 009998

Contents

Overview		v
Chapter 1	Functions	1
Chapter 2	Linear Functions	5
Chapter 4	Composite Functions and Inverses	22
Chapter 6	Quadratic Functions	25
Chapter 7	Systems of Linear Equations and Matrices	41
Chapter 10	Exponential and Logarithmic Functions	51
Chapter 12	Geometric Series	68
Chapter 17	Linear Inequalities in Two Variables and Systems of Inequalities	70

Acknowledgments

We wish to thank our colleagues and staff at Kennesaw State University. In particular, we acknowledge Sherri Delay and Arlean Paige for their help with the manuscript preparation, Tonya Jones for her encouragement and many helpful suggestions, and Ron Biggers for his support.

We also wish to express thanks to Joe Vetere for his technology assistance, to Donna Bagdasarian for her patience and editorial assistance, and to Kurt Norlin for carefully proofreading the manuscript.

Overview

This manual contains solutions to the odd numbered Things To Do in "prerequisite" chapters (Chapters covering mathematical concepts) in the text *Earth Algebra: College Algebra with Applications to Environmental Issues*, Second Edition. The chapters included are Chapters 1, 2, 4, 6, 7, 10, 12, and 17. Equations for models and answers to group work studies are not included.

This work was supported by NSF Grant USE - 9150624 and U. S. Department of Education: FIPSE Grant P116B1-61.

STUDENT'S SOLUTIONS MANUAL

to accompany

EARTH ALGEBRA
SECOND EDITION

Chapter 1 Functions

Section 1.5
Part A
1. Ordered Pairs: $\{(0,23),(1,13),(2,7),(3,5),(4,7),(5,13),(6,23)\}$

Table of Values: 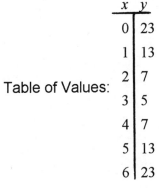 Arrow Diagram: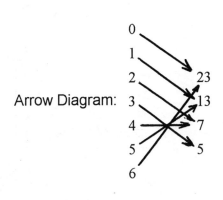

Graph:

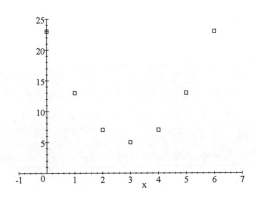

3. Ordered Pairs: $\{(1,13),(2,12),(3,11),(4,10)\}$

Formula: $y = 14 - x$
Graph:

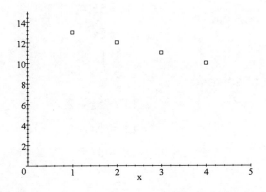

Arrow Diagram:
$$1 \to 13$$
$$2 \to 12$$
$$3 \to 11$$
$$4 \to 10$$

5. Ordered Pairs: $\{(0,1),(1,-1),(2,1),(3,-1),(4,1),(5,-1),(6,1)\}$

Arrow Diagram:

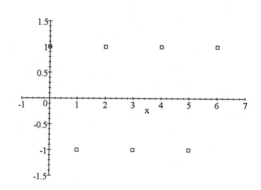

Table of Values:

x	y
0	1
1	-1
2	1
3	-1
4	1
5	-1
6	1

Graph:

7. a. $f(0) = 5(0) - 2 = -2$
b. $f(2) = 5(2) - 2 = 8$
c. Replace $f(x)$ with 3 and solve for x:
$$3 = 5x - 2$$
$$5 = 5x$$
$$x = 1.$$

d. Replace $f(x)$ with 13 and solve for x:
$$13 = 5x - 2$$
$$15 = 5x$$
$$x = 3.$$

e. Replace both x and $f(x)$ with a and solve for a :
$$a = 5a - 2$$
$$-4a = -2$$
$$a = \tfrac{1}{2}.$$

9. a. $f(2\tfrac{1}{2}) = 0$
   b. $f(1\tfrac{1}{2}) = 2$ and $f(6\tfrac{1}{2}) = 2$
   c. $f(2) + f(4) = 1 + (-1) = 0$
   d. $f(2+4) = f(6) = 1$
   e. $f(f(6)) = f(1) = 3$
   f. $f(1.8) = 1.8$

11. a. Not a function because the two points $(2,3)$ and $(2,6)$ have the same first coordinate.
    Domain: $\{0,1,2,3,4\}$; Range: $\{0,3,6\}$
    b. A function by the definition.
    Domain: $\{0,2,4,6,8\}$; Range: $\{3,6,9\}$
    c. A function by the definition.
    Domain: $\{0,1,-1,2,-2\}$; Range: $\{0,1,4\}$
    d. Not a function because the two points $(1,3)$ and $(1,7)$ have the same first coordinate.
    Domain: $\{-2,1,2,3\}$; Range: $\{1,3,5,7\}$
    e. Not a function by the vertical line test.

Part B

1. $f(2) = 1, f(-3) = 11, f(1.01) = 2.98, f(-\frac{1}{2}) = 6$
   Domain: all real numbers

3. $h(2) = 7, h(-3) = -1.1579, h(1.01) = 0.6553, h(-\frac{1}{2}) = 0.4722$
   Domain: $\{x \mid x \neq \frac{7}{4}\}$

5. $T(2.4) = -119.988, T(24) = -119.88, T(240) = -118.8, T(2400) = -108, T(24,000) = 0$
   Domain: all real numbers

7. $H(2) = -1, H(-3.3) = -1.57, H(11.001) = 32.55, H(5.5)$ is undefined
   Domain: $\{x \mid x \neq \frac{11}{2}\}$

9. $R(0) = \sqrt{63} = 7.9373, R(-2) = 8.198, R(30) = 0, R(-1.52) = 8.136,$
   $R(50) = \sqrt{-42}$, so $R(50)$ is not a real number;
   Domain: $\{s \mid s \leq 30\}$

11. $f(2) = 12, f(-3) = 47, f(1.01) = 14.9601, f(\frac{1}{2}) = 17.25$
    Domain: all real numbers

13. $K(0) = 18.2, K(-3) = -20.8857, K(200) = 2414.5910, K(2000) = 24,014.599$
    Domain: $\{t \mid t \neq \frac{1}{2}\}$.

15. $Q(0.001) = 0.001506, Q(-0.001) = 0.001510, Q(0.5) = 188.5,$
    $Q(\frac{3}{7}) = 158.2741$
    Domain: all real numbers

Chapter 2 Linear Functions

Section 2.5

1. $y = 1.4x$; slope $= 1.4$; y-intercept $= (0,0)$.

3. $y + 1.3 = -2(x - 1.1)$
   $y + 1.3 = -2x + 2.2$
   $y = -2x + 0.9$;
   slope $= -2$; y-intercept $= (0, 0.9)$.

5. $3x + 5y = 18$
   $5y = -3x + 18$
   $y = -\frac{3}{5}x + \frac{18}{5}$;
   slope $= -\frac{3}{5}$; y-intercept $= \left(0, \frac{18}{5}\right)$.

7. $12(2x - 1) - 5(3y + 2) = 8$
   $24x - 12 - 15y - 10 = 8$
   $-15y = -24x + 12 + 10 + 8$
   $-15y = -24x + 30$
   $y = \frac{8}{5}x - 2$;
   slope $= \frac{8}{5}$; y-intercept $= (0, -2)$.

9. $2(2x + y) - 4(x + y) = 6$
   $4x + 2y - 4x - 4y = 6$
   $-2y = 6$
   $y = -3$;
   slope $= 0$; y-intercept $= (0, -3)$.

11. $2.3y = 1.4x - 5$
    $y = \frac{1.4}{2.3}x - \frac{5}{2.3}$
    $y = 0.6087x - 2.1739$;
    slope $= 0.6087$; y-intercept $= (0, -2.1739)$.

13. $8(x - 1) = 9(y + 3)$
    $8x - 8 = 9y + 27$
    $8x - 35 = 9y$
    $y = \frac{8}{9}x - \frac{35}{9}$;
    slope $= \frac{8}{9}$; y-intercept $= \left(0, -\frac{35}{9}\right)$.

15. $\frac{5x+4y}{3} - \frac{y}{5} = 7$
    $15\left(\frac{5x+4y}{3} - \frac{y}{5}\right) = 15(7)$
    $5(5x + 4y) - 3(y) = 105$

$25x + 20y - 3y = 105$
$25x + 17y = 105$
$17y = -25x + 105$
$y = -\frac{25}{17}x + \frac{105}{17}$;
slope $= -\frac{25}{17}$; y-intercept $= \left(0, \frac{105}{17}\right)$.

17. Begin at point $(2, -3)$; move one unit to the right and 4 units up. The line passes through $(2, -3)$ and $(3, 1)$.

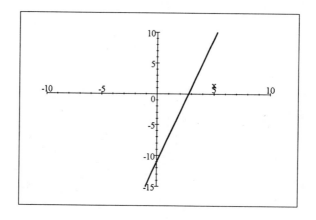

19. Begin at the point $(2, 5)$; move 5 units to the right and 3 units down. The line passes through the points $(2, 5)$ and $(7, 2)$.

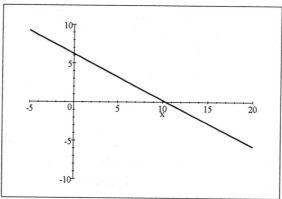

21. Since the slope is 0, this is a horizontal line through the point $(2.3, 5)$.

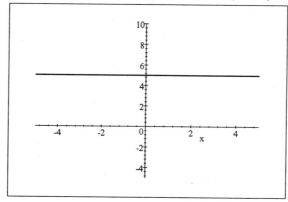

23. Begin at the point (−3,−1); move 1 unit to the right and 1 unit down. The line passes through the points (−3,−1) and (−2,−2),

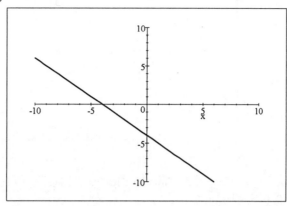

25. Begin at the point (0,0); move 2 units to the right and 1 unit down. The line passes through the points (0,0) and (2,−1),

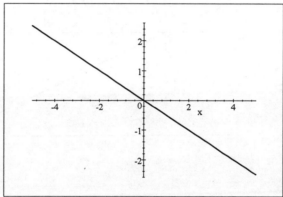

27. Since the slope is 0, this is a horizontal line passing through the point (3,−1).

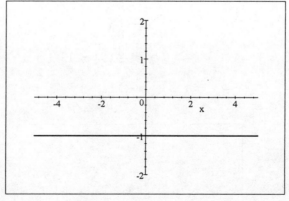

29. Begin at the point (0.5,−0.3); move 3 units to the right and 2 units down. The line passes through the points (0.5,−0.3) and (3.5,−2.3).

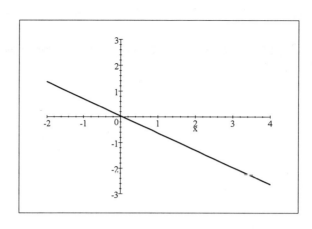

31. Begin at the point $(10,-7)$; move 10 units to the right and 7 units down. The line passes through the points $(10,-7)$ and $(20,-14)$.

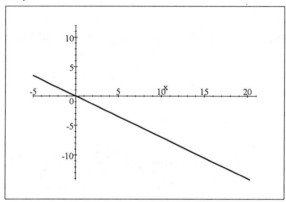

33. $y = \frac{3}{8}x - 5$
    x-intercept: $(13.3, 0)$

$$0 = \frac{3}{8}x - 5$$
$$5 = \frac{3}{8}x$$
$$x = \frac{8}{3}(5) = 13.3$$

y-intercept: $(0,-5)$
slope: $\frac{3}{8}$

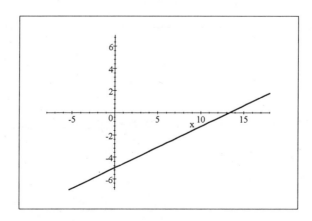

8

35. $y + 5 = -\frac{3}{4}(x - 6)$
$y = -\frac{3}{4}x - \frac{1}{2}$
x-intercept: $\left(-\frac{2}{3}, 0\right)$

$$0 = -\frac{3}{4}x - \frac{1}{2}$$
$$\frac{1}{2} = -\frac{3}{4}x$$
$$x = -\frac{4}{3}\left(\frac{1}{2}\right) = -\frac{2}{3}$$

y-intercept: $\left(0, -\frac{1}{2}\right)$
slope: $-\frac{3}{4}$

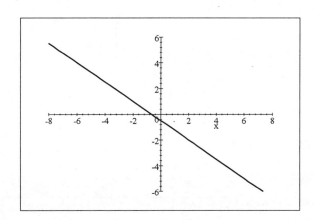

37. $x = 5$
This is the equation of a vertical line.
x-intercept: $(5, 0)$
y-intercept: none
slope: undefined

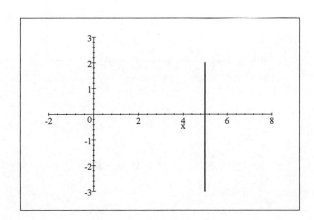

39. $y = 1 - 4x = -4x + 1$
x-intercept: $\left(\frac{1}{4}, 0\right)$

$$0 = -4x + 1$$
$$4x = 1$$
$$x = \frac{1}{4}$$

y-intercept: $(0, 1)$
slope: -4.

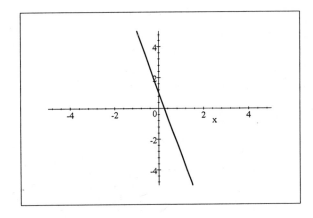

41. $x = 7$

Vertical line with **x**-intercept $(7, 0)$;
y-intercept: none
slope: undefined

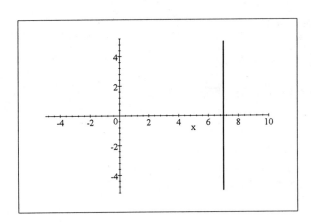

43. $y = -3x$

x-intercept: $(0, 0)$

$$0 = -3x$$
$$x = 0$$

y-intercept: $(0, 0)$
slope: -3

10

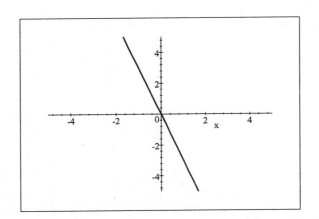

45. $x + 3y = -2x + 3y - 5$
$3x = -5$
$x = -\frac{5}{3}$
Vertical line with x-intercept $(-\frac{5}{3}, 0)$
y-intercept: none
slope: undefined.

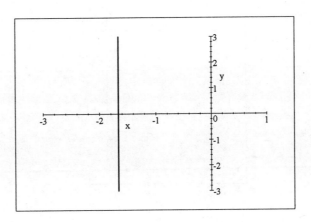

47. $y = -2.7x + 4$
x-intercept: $(1.481, 0)$

$$0 = -2.7x + 4$$
$$2.7x = 4$$
$$x = 1.481.$$

y-intercept: $(0, 4)$
slope: -2.7

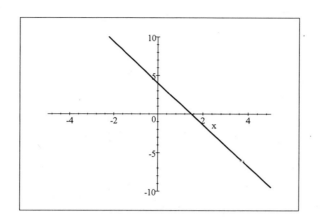

49. $y - 9 = 7$
$y = 16$
Horizontal line with y-intercept $(0, 16)$
x-intercept: none
slope: 0.

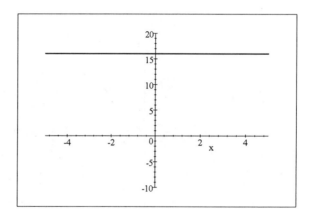

51. $x - y + 7 = 4x - 2y + 5$
$-3x + y = -2$
$y = 3x - 2$
x-intercept: $(\frac{2}{3}, 0)$

$$0 = 3x - 2$$
$$-3x = -2$$
$$x = \frac{2}{3}.$$

y-intercept: $(0, -2)$
slope: 3.

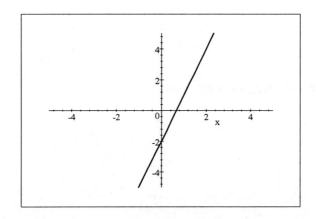

Section 2.6

1. slope $=\frac{1-5}{2-3} = 4$;
   using the point-slope formula with the point $(3, 5)$,
   $$y - 5 = 4(x - 3)$$
   $$y - 5 = 4x - 12$$
   $$y = 4x - 7.$$

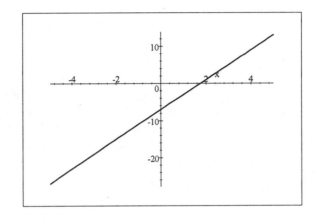

3. slope $=\frac{1-1}{-2-3} = 0$;
   using the point-slope formula with the point $(3, 1)$,
   $$y - 1 = 0(x - 3)$$
   $$y - 1 = 0$$
   $$y = 1.$$

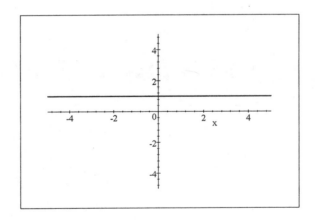

5. Using the point-slope formula,
   $$y - 2 = \frac{1}{2}(x + 3)$$
   $$y - 2 = \frac{1}{2}x + \frac{3}{2}$$
   $$y = \frac{1}{2}x + \frac{7}{2}.$$

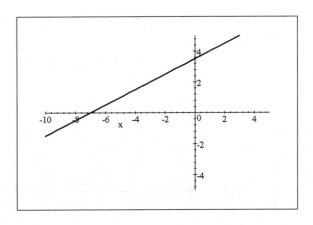

7. Using the point-slope formula,
$$y + 12 = 16.4(x - 0)$$
$$y + 12 = 16.4x$$
$$y = 16.4x - 12$$

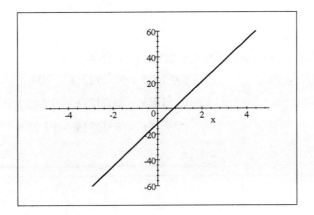

9. slope $= \frac{350-280}{51-(-30)} = 0.8642$.

using the point-slope formula with the point $(-30, 280)$,
$$y - 280 = 0.8642(x - (-30))$$
$$y - 280 = 0.8642x + 25.926$$
$$y = 0.8642x + 305.93$$

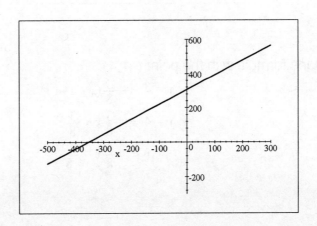

11. Using the point-slope formula,
$$y - 2.4 = 0(x - 12)$$
$$y - 2.4 = 0$$
$$y = 2.4.$$

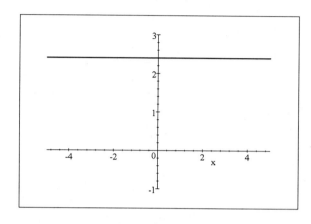

13. slope $= \frac{19.4 - 18.5}{48 - 20} = 0.0321$

    using the point-slope formula with the point $(20, 18.5)$,
$$y - 18.5 = 0.0321(x - 20)$$
$$y - 18.5 = 0.0321x - 0.642$$
$$y = 0.0321x + 17.858$$

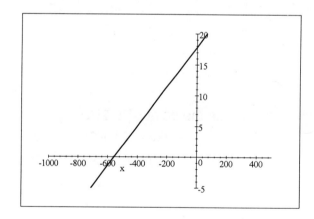

15. slope $= \frac{1-4}{7-(-1)} = -\frac{3}{8}$;

    using the point-slope formula with the point $(-1, 4)$,
$$y - 4 = -\frac{3}{8}(x - (-1))$$
$$y - 4 = -\frac{3}{8}x - \frac{3}{8}$$
$$y = -\frac{3}{8}x + \frac{29}{8}$$

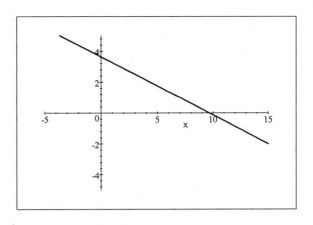

17. Since the slope is undefined, the line is vertical.

The equation is $x = -1$.

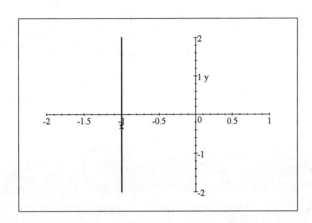

19. slope $= \frac{34,000 - 10,000}{5-3} = 12,000$.

Using the point-slope formula with the point (3, 10000),
$$y - 10,000 = 12,000(x - 3)$$
$$y - 10,000 = 12,000x - 36,000$$
$$y = 12,000x - 26,000.$$

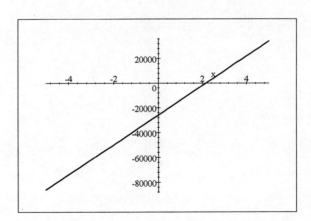

17

21. This is a horizontal line (since the slope is 0) with equation $y = 0$.

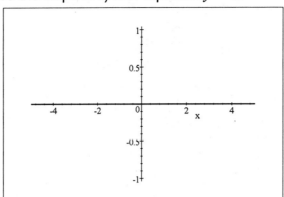

23. Replace $F(C)$ with C and solve for C:

$C = \frac{9}{5}C + 32$

$-\frac{4}{5}C = 32$

$C = \left(-\frac{5}{4}\right)32 = -40°$.

Section 2.7

1. $y = 2.4x + 12$
   The slope is $m = 2.4$.

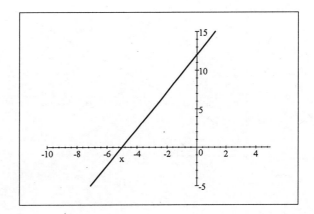

3. $CO(t) = 0.0003t + 12$
   The slope is $m = 0.0003$.
   Window $[-50,000, 10,000, 5000]$ by $[-10, 20, 2]$

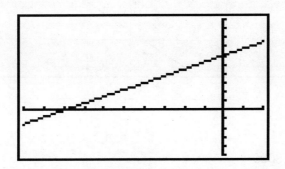

5. $m(t) = 2(3t - 5) + 0.1(1 - 2t) = 5.8t - 9.9$
   The slope is $m = 5.8$.

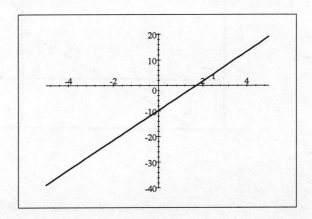

7. $E(t) = 0.01t + 115.2$
The slope is $m = 0.01$.

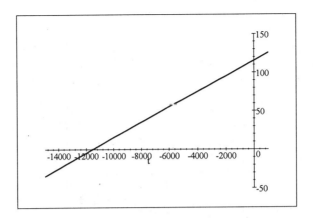

9. $f(x) = 1.6x - 4$
The slope is $m = 1.6$.

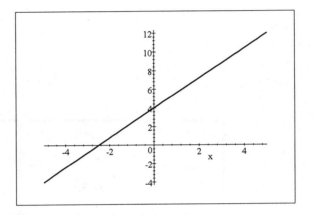

11. $s(t) = 0.0005t + 11$
The slope is $m = 0.0005$.
Window $[-50,000, 10,000, 5000]$ by $[-10, 20, 2]$

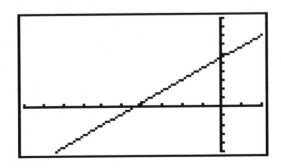

13. $B(t) = \frac{3\left(t-\frac{1}{2}\right)}{7} + \frac{4(6-t)}{5} = \frac{3t-\frac{3}{2}}{7} + \frac{24-4t}{5} = \frac{6t-3}{14} + \frac{24-4t}{5} = \frac{30t-15+336-56t}{70} = \frac{-26t+321}{70} = -\frac{13}{35}t + \frac{321}{70} = -0.37143t + 4.5857.$

The slope is $m = -\frac{13}{35} = -0.37143$.

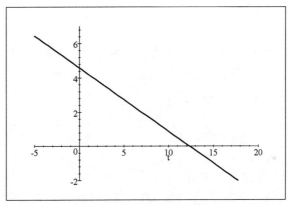

Chapter 4 Composite Functions and Inverses

Section 4.2

1. $x \xrightarrow{M_3} 3x \xrightarrow{S_2} (3x-2) \xrightarrow{D_4} \frac{3x-2}{4}$

3. $x \xrightarrow{D_4} \frac{x}{4} \xrightarrow{A_7} \left(7+\frac{x}{4}\right) \xrightarrow{M_3} 3\left(7+\frac{x}{4}\right) \xrightarrow{S_5} 3\left(7+\frac{x}{4}\right)-5$

5. $x \xrightarrow{M_3} 3x \xrightarrow{S_2} (3x-2) \xrightarrow{D_4} \frac{7}{3x-2}$

   $10 \xrightarrow{D_3} 30 \xrightarrow{A_2} 28 \xrightarrow{M_4} 7$

   Thus, the solution is $x = 10$.

7. $x \xrightarrow{D_4} \frac{x}{4} \xrightarrow{A_7} \left(7+\frac{x}{4}\right) \xrightarrow{M_3} 3\left(7+\frac{x}{4}\right) \xrightarrow{S_5} 3\left(7+\frac{x}{4}\right)-5$

   $8 \xrightarrow{M_4} 2 \xrightarrow{S_7} 9 \xrightarrow{D_3} 27 \xrightarrow{A_5} 22$

   Thus, the solution is $x = 8$.

9. a. $m = \frac{212-32}{100-0} = \frac{9}{5}$

   Use the point slope formula with the point $(0, 32)$,

   $$F - 32 = \frac{9}{5}(C - 0)$$
   $$F - 32 = \frac{9}{5}C$$
   $$F = \frac{9}{5}C + 32.$$

   In function notation, $F(C) = \frac{9}{5}C + 32 = 1.8C + 32$.

   b.
   $$F = \frac{9}{5}C + 32$$
   $$F - 32 = \frac{9}{5}C$$
   $$5(F - 32) = 9C$$
   $$C = \frac{5(F-32)}{9} = 0.55556F - 17.778.$$

11. $H[(f(x)] = H[5(2-3x)] = 3 + (2(5(2-3x)) - 5) = 3 + 10(2-3x) - 5 = 3 + 20 - 30x - 5 = 18 - 30x$
    $H[f(0.2)] = 18 - 30(0.2) = 12.$

13. $AB(-2) = -2 + 2 = 0$;
    $C[AB(-2)] = C(0) = -2$

15. $PQ[QP(x)] = PQ\left(\frac{x-1.5}{3}\right) = 3\left(\frac{x-1.5}{3}\right) + 1.5 = x - 1.5 + 1.5 = x.$

17. $A[R(T(v))] = A[R(\sqrt{v}\,)] = A\left(\sqrt{v}^2 + 1\right) = A(v+1) = v+1;$
$R[A(T(v))] = R[A(\sqrt{v}\,)] = R(\sqrt{v}) = \sqrt{v}^2 + 1 = v+1;$
$T(A(R(u))) = T[A(u^2+1)] = T(u^2+1) = \sqrt{u^2+1}.$

19. $O(M(t)) = O\left(\frac{t^2}{3}\right) = \frac{2}{\frac{t^2}{3}} = 2\left(\frac{3}{t^2}\right) = \frac{6}{t^2};$
$M(O(3)) = M\left(\frac{2}{3}\right) = \frac{\left(\frac{2}{3}\right)^2}{3} = \frac{\frac{4}{9}}{3} = \frac{4}{9} \cdot \frac{1}{3} = \frac{4}{27}.$

21. $A(D(r)) = A\left(\sqrt{2r^2-3}\right) = \sqrt{\frac{2r^2-3+3}{2}} = \sqrt{r^2} = |r|;$
$A(D(1000)) = 1000.$

23. $N(A(P(u))) = N[A\left(\frac{u^2}{4}\right)] = N\left(\left(\frac{u^2}{4}\right)^3\right) = N\left(\frac{u^6}{64}\right) = \sqrt{\frac{u^6}{64} - 1} = \frac{1}{8}\sqrt{(u^6 - 64)};$
$P(A(N(s))) = P\left(A\left(\sqrt{s-1}\right)\right) = P\left(\left(\sqrt{s-1}\right)^3\right) = \frac{(s-1)^3}{4}, s \geq 1.$

25. a. $P(F(w)) = P(50 - 0.5w) = 24 - 0.4(50 - 0.5w) = 4 + 0.2w.$

b. $P(0) = 4 + 0.2(0) = \$4.00.$

c. $4 + 0.2w = 12$
$0.2w = 8$
$w = 40$ gallons per day

d. $50 - 0.5w = 0$
$50 = 0.5w$
$w = 100$ gallons per day

Section 4.3

1. $10^{0.5} = 3.162$

3. $10^{-1.36} = 0.044$

5. $e^{4.7} = 109.947$

7. $\log(3.162) = 0.500$

9. $\log(0.0437) = -1.360$

11. $\ln(0.05) = -2.996$

13.

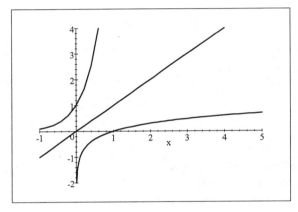

$y = 10^x, y = x, y = \log x$

The graph of $y = \log x$ is the reflection of the graph of $y = 10^x$ across the line $y = x$.

CHAPTER 6 Quadratic Functions

Section 6.2

1. $a = 1 > 0$ so the graph opens up;
   to find the y-intercept, replace x with 0
   $$y = 0^2 - 6(0) + 5 = 5$$
   so the y-intercept is $(0, 5)$;
   to find the x-intercepts replace y with 0 and solve for x,
   $$0 = x^2 - 6x + 5$$
   $$0 = (x - 5)(x - 1)$$
   $$x = 1 \text{ or } x = 5;$$
   the x-intercepts are $(1, 0)$ and $(5, 0)$.

3. $a = 0.02 > 0$ so the graph opens up;
   y-intercept: $(0, 27)$
   $$y = 0.02(0)^2 - 63.1(0) + 27 = 27$$
   x-intercepts: $(3154.57, 0)$ and $(0.43, 0)$
   $$0 = 0.02x^2 - 63.1x + 27$$
   $$x = \frac{63.1 \pm \sqrt{(-63.1)^2 - 4(0.02)(27)}}{2(0.02)}$$
   $$x = 3154.57 \text{ or } x = 0.43$$

5. $a = -3 < 0$ so the graph opens down;
   E-intercept: $(0, 13.1)$
   $$E(0) = -3(0)^2 + 27.5(0) + 13.1 = 13.1$$
   z-intercepts: $(-0.45, 0)$ and $(9.62, 0)$
   $$0 = -3z^2 + 27.5z + 13.1$$
   $$z = \frac{-27.5 \pm \sqrt{(27.5)^2 - 4(-3)(13.1)}}{2(-3)}$$
   $$z = -0.45 \text{ or } z = 9.62.$$

7. $a = -4 < 0$ so the graph opens down;
   y-intercept: $(0, 9)$
   $$G(0) = 9 - 4(0)^2 + 0.6(0) = 9$$
   x-intercepts: $(-1.43, 0)$ and $(1.58, 0)$
   $$0 = 9 - 4x^2 + 0.6x,$$
   write the equation in standard form
   $$0 = -4x^2 + 0.6x + 9$$
   and solve using the quadratic formula with $a = -4, b = 0.6$, and $c = 9$.

9. $a = -3.5 < 0$ so the graph opens down;
   y-intercept: $(0,0)$
   $$A(0) = 0 - 3.5(0)^2 = 0$$
   x-intercepts: $(0,0)$ and $(0.2857, 0)$
   $$0 = x - 3.5x^2,$$
   write the equation in standard form
   $$0 = -3.5x^2 + x$$
   and solve using the quadratic formula with $a = -3.5, b = 1$, and $c = 0$.

11. $a = 7.14 > 0$ so the graph opens up;
    y-intercept: $(0, 3)$
    $$C(0) = 7.14(0)^2 + 3 = 3$$
    x-intercepts: none
    $$0 = 7.14x^2 + 3,$$
    solve using the quadratic formula with $a = 7.14, b = 0$, and $c = 3$; the equation has no real solutions so there are no x-intercepts.

13. $f(x) = 3(x-4)^2 - 7 = 3x^2 - 24x + 41$
    $a = 3 > 0$ so the graph opens up;
    y-intercept: $(0, 41)$
    $$f(0) = 3(0)^2 - 24(0) + 41 = 41$$
    x-intercepts: $(5.528, 0)$ and $(2.472, 0)$
    $$0 = 3x^2 - 24x + 41,$$
    solve using the quadratic formula, with $a = 3, b = -24, c = 41$.

15. $a = -4 < 0$ so the graph opens down;
    y-intercept: $(0, 7)$
    $$y = 7 + 3(0) - 4(0)^2 = 7,$$
    x-intercepts: $(-1, 0)$ and $\left(\frac{7}{4}, 0\right)$
    $$0 = 7 + 3x - 4x^2$$
    solve using the quadratic formula with $a = -4, b = 3, c = 7$;

17. $a = 6.5 > 0$ so the graph opens up;
    y-intercept: $(0, 7.6)$
    $$K(0) = 6.5(0)^2 - 3.1(0) + 7.6 = 7.6$$
    x-intercepts: none
    $$0 = 6.5x^2 - 3.1x + 7.6$$
    Using the quadratic formula with $a = 6.5, b = -3.1, c = 7.6$, we find that the function has no real solutions so there are no x-intercepts.

19. $N(x) = (10x + 3.4)^2 = 100x^2 + 68x + 11.56$
$a = 100 > 0$ so the graph opens up;
y-intercept: $(0, 11.56)$

$$N(0) = 100(0)^2 + 68(0) + 11.56 = 11.56$$

x-intercept: $(-0.34, 0)$

$$0 = 100x^2 + 68x + 11.56$$

Solve using the quadratic formula with $a = 100, b = 68, c = 11.56$. The solution is -0.34.

Section 6.3

1. The x-coordinate of the vertex is $\frac{-b}{2a} = \frac{-6.2}{2(1)} = -3.1$;
   the y-coordinate is $y = (-3.1)^2 + 6.2(-3.1) + 15.7 = 6.09$.
   Thus the vertex is $(-3.1, 6.09)$.

3. The x-coordinate of the vertex is $\frac{-b}{2a} = \frac{21.4}{2(33)} = 0.324$;
   the y-coordinate is $y = 97.1 - 21.4(0.324) + 33(0.324)^2 = 93.631$.
   Thus the vertex is $(0.324, 93.631)$.

5. The x-coordinate of the vertex is $\frac{-b}{2a} = \frac{-6}{2(4)} = \frac{-3}{4} = -0.75$;
   the y-coordinate is $y = f(-0.75) = 4(-0.75)^2 + 6(-0.75) + 8 = 5.75$.
   Thus the vertex is $(-0.75, 5.75)$.

7. The x-coordinate of the vertex is $\frac{-b}{2a} = \frac{0}{2(352.1)} = 0$;
   the y-coordinate is $y = f(0) = 352.1(0)^2 = 0$.
   Thus the vertex is $(0, 0)$.

9. The x-coordinate of the vertex is $\frac{-b}{2a} = \frac{1}{2(4)} = \frac{1}{8} = 0.125$;
   the y-coordinate is $y = A(0.125) = -0.0625$.
   Thus the vertex is $(0.125, -0.0625)$.

11. The x-coordinate of the vertex is $\frac{-b}{2a} = \frac{-6}{2(-1)} = 3$;
    the y-coordinate is $y = g(3) = 12$.
    Thus the vertex is $(3, 12)$.

13. First, write the equation in standard form,
    $$i(x) = x(x-4) + 3 = x^2 - 4x + 3.$$
    The x-coordinate of the vertex is $\frac{-b}{2a} = \frac{4}{2(1)} = 2$;
    the y-coordinate is $y = i(2) = -1$.
    Thus the vertex is $(2, -1)$.

15. This equation is given in vertex form, so the vertex is $(4, -7)$.

17. First, write the equation in standard form,
    $$m(x) = (2.7x - 1.4)(3.5x + 0.2) = 9.45x^2 - 4.36x - 0.28.$$
    The x-coordinate of the vertex is $\frac{-b}{2a} = \frac{4.36}{2(9.45)} = 0.231$;
    the y-coordinate is $y = m(0.231) = -0.783$.
    Thus the vertex is $(0.231, -0.783)$.

Section 6.4

The maximum or minimum value is the y-coordinate of the vertex, so we first find the vertex and then determine whether the parabola opens up or down.

1. $f(x) = x^2 - 4x + 3$
   The x-coordinate of the vertex is $x = \frac{4}{2} = 2$;
   the y-coordinate is $y = f(2) = 2^2 - 4(2) + 3 = -1$.
   The vertex is $(2, -1)$ and the graph opens up since $a = 1 > 0$;
   minimum: -1, range: $y \geq -1$

3. $H(t) = 110 + 4t^2 = 4t^2 + 110$
   The t-coordinate of the vertex is $t = \frac{0}{2(4)} = 0$;
   the H-coordinate is $H(0) = 4(0)^2 + 110$.
   The vertex is $(0, 110)$ and the graph opens up since $a = 4 > 0$
   minimum: 110, range: $y \geq 110$

5. $T(x) = 14x^2$
   The vertex is $(0, 0)$ and the graph opens up since $a = 14 > 0$
   minimum: 0, range $y \geq 0$

7. $G(t) = t - 15 - 2.1t^2 = -2.1t^2 + t - 15$
   The t-coordinate of the vertex is $t = \frac{-1}{2(-2.1)} = 0.2381$;
   the G-coordinate is $G(0.2381) = -2.1(0.2381)^2 + (0.2381) - 15 = -14.881$.
   The vertex is $(0.2381, -14.881)$ and the graph opens down since $a = -2.1 < 0$.
   maximum: -14.881, range: $y \leq -14.881$.

9. $g(t) = 0.5t^2 - t + 2$
   The t-coordinate of the vertex is $t = \frac{1}{2(0.5)} = 1$;
   the g-coordinate is $g(1) = 0.5(1)^2 - 1 + 2 = 1.5$.
   The vertex is $(1, 1.5)$ and the graph opens up since $a = 0.5 > 0$
   minimum: 1.5, range: $y \geq 1.5$

11. $l(x) = 37 - x^2 = -x^2 + 37$
    The x-coordinate of the vertex is $\frac{0}{2(-1)} = 0$;
    the l-coordinate is $l(0) = 37$.
    The vertex is $(0, 37)$ and the graph opens down since $a = -1 < 0$
    maximum: 37, range: $y \leq 37$

13. $O(s) = (3 - s)(4 + s) = -s^2 - s + 12$
    The s-coordinate of the vertex is $s = \frac{1}{2(-1)} = -\frac{1}{2}$;
    the O-coordinate is $O\left(-\frac{1}{2}\right) = -\left(-\frac{1}{2}\right)^2 - \left(-\frac{1}{2}\right) + 12 = \frac{49}{4} = 12.25$.
    The vertex is $(-0.5, 12.25)$ and the graph opens down since $a = -1 < 0$
    maximum: 12.25, range: $y \leq 12.25$

15. $Q(x) = 3(x-4)^2 - 7 = 3x^2 - 24x + 41$
    The vertex is $(4, -7)$ and the graph opens up since $a = 3 > 0$
    minimum: -7, range: $y \geq -7$.

Section 6.5

1. $y = x^2 - 6x - 10$

y-intercept: $(0, -10)$

$$y = (0)^2 - 6(0) - 10 = -10$$

x-intercepts: $(7.36, 0)$ and $(-1.36, 0)$

$$0 = x^2 - 6x - 10$$
$$x = \frac{-(-6) \pm \sqrt{(-6)^2 - 4(1)(-10)}}{2(1)} = 7.36, -1.36.$$

The x-coordinate of the vertex is $x = \frac{6}{2(1)} = 3$;
the y-coordinate is $y = (3)^2 - 6(3) - 10 = -19$
so the vertex is $(3, -19)$.
Since $a = 1 > 0$, the parabola opens up. The minimum value of the function is -19 and the range is $y \geq -19$.

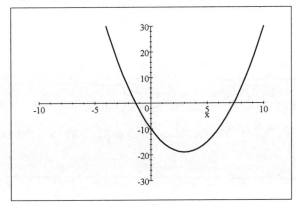

3. $y = 4 - 6x^2 = -6x^2 + 4$

y-intercept: $(0, 4)$

$$y = -6(0)^2 + 4 = 4$$

x-intercepts: $(0.82, 0)$ and $(-0.82, 0)$

$$0 = -6x^2 + 4$$
$$6x^2 = 4$$
$$x^2 = \frac{2}{3}$$
$$x = \pm\sqrt{\frac{2}{3}} = \pm 0.82.$$

The x-coordinate of the vertex is $\frac{-0}{2(-6)} = 0$;
the y-coordinate is $y = -6(0)^2 + 4 = 4$
so the vertex is $(0, 4)$.
Since $a = -6 < 0$, the parabola opens down. The maximum value is 4 and the range is $y \leq 4$.

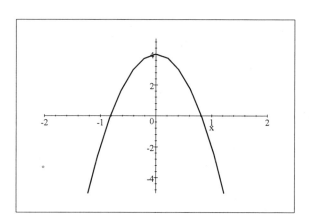

5. $y = 2.5x^2 - x + 7.3$
   y-intercept: $(0, 7.3)$

$$y = 2.5(0)^2 - 0 + 7.3 = 7.3.$$

x-intercepts: none

$$0 = 2.5x^2 - x + 7.3$$
$$x = \frac{-(-1) \pm \sqrt{(-1)^2 - 4(2.5)(7.3)}}{2(2.5)} = \frac{-(-1) \pm \sqrt{-72}}{2(2.5)}.$$

Since the square root of a negative number is not a real number, the equation has no solution.
The x-coordinate of the vertex is $x = \frac{1}{2(2.5)} = 0.2$;
the y-coordinate is $y = 2.5(0.2)^2 - 0.2 + 7.3 = 7.2$
so the vertex is $(0.2, 7.2)$.
Since $a = 2.5 > 0$, the parabola opens up. The minimum value of the function is 7.2 and the range is $y \geq 7.2$.

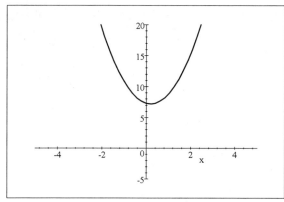

7. $A(t) = 0.1t^2 - 0.14t$
   A-intercept: $(0, 0)$;

$$A(0) = 0.01(0)^2 - 0.14(0) = 0$$

32

t-intercepts: $(0, 0)$ and $(14, 0)$

$$0 = 0.01t^2 - 0.14t$$
$$0 = t(0.01t - 0.14)$$
$$t = 0 \text{ or } 0.01t - 0.14 = 0$$
$$t = 0 \text{ or } t = 14$$

The t-coordinate of the vertex is $t = \frac{0.14}{2(0.01)} = 7$; the A-coordinate is $A(7) = 0.01(7)^2 - 0.14(7) = -0.49$ so the vertex is $(7, -0.49)$.

Since $a = 0.01 > 0$, the parabola opens up. The minimum value of the function is -0.49 and the range is $y \geq -0.49$.

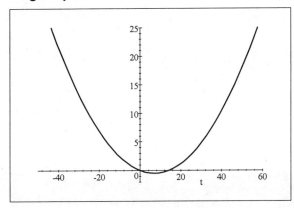

9. $S(t) = 2t - t^2 - 1 = -t^2 + 2t - 1$

S-intercept: $(0, -1)$

$$S(0) = -(0)^2 + 2(0) - 1 = -1$$

t-intercept: $(1, 0)$ and

$$0 = -t^2 + 2t - 1$$
$$0 = t^2 - 2t + 1$$
$$0 = (t - 1)^2$$
$$t = 1.$$

The t-coordinate of the vertex is $t = \frac{-2}{2(-1)} = 1$; the S-coordinate is $S(1) = -(1)^2 + 2(1) - 1 = 0$, so the vertex is $(1, 0)$.

Since $a = -1 < 0$, the parabola opens down. The maximum value of the function is 0 and the range is $y \leq 0$.

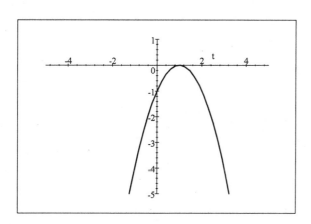

11. $T(x) = (0.24x - 1.64)(x - 3) = 0.24x^2 - 2.36x + 4.92$
    T-intercept: $(0, 4.92)$

    $$T(0) = 0.24(0)^2 - 2.36(0) + 4.92 = 4.92$$

    x-intercepts: $(6.83, 0)$ and $(3, 0)$

    $$0 = (0.24x - 1.64)(x - 3)$$
    $$0.24x - 1.64 = 0 \text{ or } x - 3 = 0$$
    $$x = 6.83 \text{ or } x = 3.$$

    The x-coordinate of the vertex is $x = \frac{2.36}{2(0.24)} = 4.92$;
    the T-coordinate is $T(4.92) = 0.24(4.92)^2 - 2.36(4.92) + 4.92 = -0.88$
    so the vertex is $(4.92, -0.88)$.
    Since $a = 0.24 > 0$, the parabola opens up. The minimum value of the function is -0.88, and the range is $y \geq -0.88$

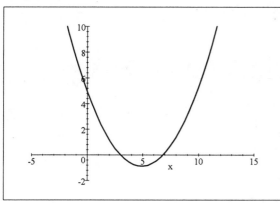

13. $p(s) = 84s(300s - 27) = 25\,200s^2 - 2268s$
    p-intercept: $(0, 0)$

    $$p(0) = 25200(0)^2 - 2268(0) = 0$$

    s-intercepts: $(0, 0)$ and $(0.09, 0)$

34

$$0 = 84s(300s - 27)$$
$$84s = 0 \text{ or } 300s - 27 = 0$$
$$s = 0 \text{ or } s = 0.09.$$

The s-coordinate of the vertex is $s = \frac{2268}{2(25200)} = 0.045$;
the p-coordinate is $p(0.045) = 25{,}200(0.045)^2 - 2268(0.045) = -51.03$, so the vertex is $(0.045, -51.03)$.

Since $a = 25{,}200 > 0$, the parabola opens up. The minimum value of the function is -51.03 and the range is $y \geq -51.03$.

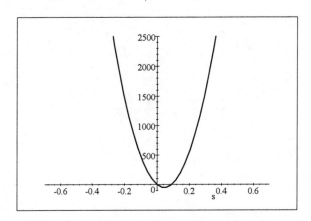

15. $y(t) = 3(t-4)^2 - 7 = 3t^2 - 24t + 41$
y-intercept: $(0, 41)$

$$y(0) = 3(0)^2 - 24(0) + 41 = 41$$

x-intercepts: $(2.4725, 0)$ and $(5.5275, 0)$

$$0 = 3t^2 - 24t + 41$$
$$t = \frac{24 \pm \sqrt{(-24)^2 - 4(3)(41)}}{2(3)}$$
$$t = 2.4725 \text{ or } t = 5.5275.$$

Since the function is given in vertex form, we see that the vertex is $(4, -7)$.
$a = 3 > 0$, so the parabola opens up. The minimum value of the function is -7 and the range is $y \geq -7$.

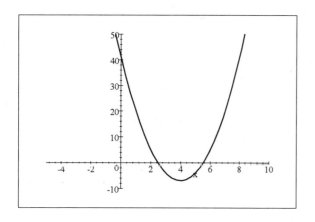

17. $12 = 14 + x - 2x^2$

In standard form the equation is

$$-2x^2 + x + 2 = 0.$$

Solve using the quadratic formula to get

$$x = -0.78, x = 1.28.$$

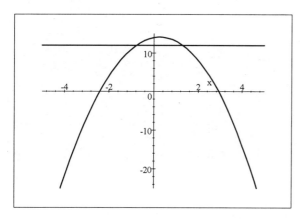

19. $2.5x^2 - x + 7.3 = -1$

In standard form the equation is

$$2.5x^2 - x + 8.3 = 0.$$

Using the quadratic formula, we find that the equation has no solution.

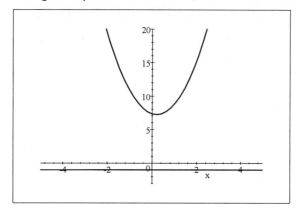

21. $15 = t^2 + 4.6t + 5.29$

In standard form the equation is
$$t^2 + 4.6t - 9.71 = 0.$$
Solve using the quadratic formula to get $t = 1.573$, $t = -6.173$

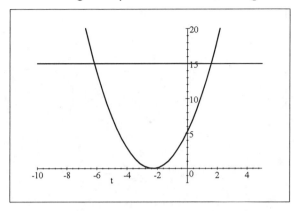

23. $375 = 84s(300s - 27)$
$375 = 25{,}200s^2 - 2268s$

In standard form the equation is
$$25{,}200s^2 - 2268s - 375 = 0.$$
Solve using the quadratic formula to get $s = 0.175$, $s = -0.085$.

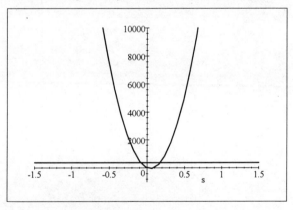

25. $3(x-4)^2 - 7 = 35$

In standard form, the equation is
$$3x^2 - 24x + 41 = 35$$
$$3x^2 - 24x + 6 = 0.$$
Solve using the quadratic formula to get $x = 7.742$, $x = 0.258$.

37

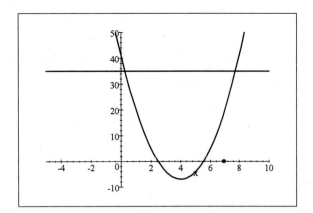

Section 6.6

1. Vertex = $(105, 6.285)$

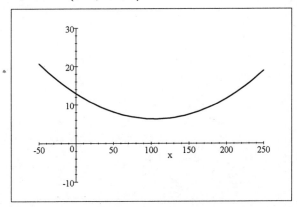

3. Vertex = $(181.67, -77.81)$
   $K(b) = 45$ when $b = 383.993$ or -20.66;
   $K(b) = 50$ when $b = 388.07$ or $b = -24.74$

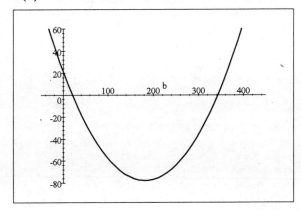

5. Vertex $(0.302, 6.559)$
   $KO(t) = 10$ when $t = -.47751$ or $t = 1.0811$
   $KO(t) = 50$ when $t = -2.4671$ or $t = 3.0707$
   $KO(t) = 100$ when $t = -3.7592$ or $t = 4.3628$

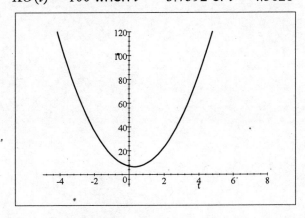

7. Vertex = $(0.5, -0.68)$
$No(s) = 10$ when $s = -1.4821$ or $s = 2.4821$
$No(s) = 20$ when $s = -2.2582$ or $s = 3.2582$
$No(s) = 0$ when $s = 0$ or $s = 1$

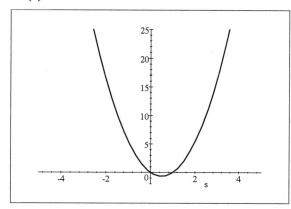

Chapter 7 Systems of Linear Equations and Matrices

Section 7.1

1. To eliminate y, add the two equations:

$$x + 3y = 1$$
$$+\ 2x - 3y = 2$$
$$3x = 3$$
$$x = 1$$

   Substitute $x = 1$ into the 1st equation to get
$$1 + 3y = 1$$
$$3y = 0$$
$$y = 0.$$

   The solution is $x = 1, y = 0$.

3. To eliminate y, multiply the first equation by 3 to get the system:
$$3x - 3y = 3$$
$$2x + 3y = -2$$

   Add the equations to get
$$5x = 1$$
$$x = \frac{1}{5}.$$

   Substitute this into the original first equation to get

$$\frac{1}{5} - y = 1$$
$$y = -\frac{4}{5}.$$

   The solution is $x = \frac{1}{5}, y = -\frac{4}{5}$ or $x = 0.2, y = -0.8$.

5. To eliminate x, multiply the second equation by -2 to get the system:
$$4x + y = 1$$
$$-4x - 6y = 4$$

   Then add the equations to get
$$-5y = 5$$
$$y = -1.$$

   Substitute this into the first equation to get
$$4x + (-1) = 1$$
$$4x = 2$$
$$x = \frac{1}{2}.$$

The solution is $x = \frac{1}{2}, y = -1$.

7. To eliminate x, multiply the first equation by -1 and add it to the second equation to get
$$-x - 2y - z = -3$$
$$+x + 3y + 4z = 6$$
$$y + 3z = 3.$$

To get another equation in y and z multiply the second equation by -2 and add it to the third equation to get
$$-2x - 6y - 8z = -12$$
$$+2x + 4y + 3z = 7$$
$$-2y - 5z = -5.$$

This gives us a system of two equations in two unknowns:
$$y + 3z = 3$$
$$-2y - 5z = -5.$$

Multiply the first equation by 2 to get the system
$$2y + 6z = 6$$
$$-2y - 5z = -5.$$

Eliminate y by adding these equations to get
$$z = 1.$$

Substitute $z = 1$ into the first two equations in the original system to get
$$x + 2y + 1 = 3$$
$$x + 3y + 4 = 6$$

which simplifies to the system
$$x + 2y = 2$$
$$x + 3y = 2$$

To eliminate x, multiply the first equation by -1 and add it to the second equation:
$$-x - 2y = -2$$
$$+x + 3y = 2$$
$$y = 0.$$

Substitute $y = 0$ and $z = 1$ into the first equation to get:
$$x + 2(0) + 1 = 3$$
$$x = 2.$$

The solution is $x = 2, y = 0, z = 1$.

9. To eliminate y, add equation 1 to equation 2 to get
$$5x + z = 4.$$

To get a second equation in x and z, add equation 1 to equation 3 to get
$$6x + 6z = 6$$

which simplifies to

$$x + z = 1.$$

This gives a system of two equations in two unknowns, x and z:

$$5x + z = 4$$
$$x + z = 1.$$

Multiply the first equation by -1 and add these two equations to get

$$-4x = -3$$
$$x = \frac{3}{4}.$$

Substitute $x = \frac{3}{4}$ into the first two equations in the original system to get

$$2\left(\frac{3}{4}\right) + y + 3z = 1$$
$$3\left(\frac{3}{4}\right) - y - 2z = 3,$$

which simplifies to the system

$$y + 3z = -\frac{1}{2}$$
$$-y - 2z = \frac{3}{4}.$$

Adding these two equation gives

$$z = \frac{1}{4}.$$

Substitute $x = \frac{3}{4}$ and $z = \frac{1}{4}$ into the original first equation to get

$$2\left(\frac{3}{4}\right) + y + 3\left(\frac{1}{4}\right) = 1$$
$$y = 1 - \frac{6}{4} - \frac{3}{4} = -\frac{5}{4}.$$

The solution is $x = \frac{3}{4}, y = -\frac{5}{4}, z = \frac{1}{4}$.

11. To eliminate x, multiply the first equation by -20. This gives the system

$$-2x + 6y = -10$$
$$2x + 7y = 4.$$

Add the equations to get

$$13y = -6$$
$$y = \frac{-6}{13}.$$

substitute $y = -\frac{6}{13}$ into the second equation in the system to get

$$2x + 7\left(-\frac{6}{13}\right) = 4$$
$$2x - \frac{42}{13} = 4$$
$$2x = \frac{94}{13}$$
$$x = \frac{47}{13}.$$

The solution is $x = \frac{47}{13}, y = -\frac{6}{13}$ or $x = 3.6154, y = -0.46154$.

13. $2x + 3y = 4$
   $ 4y - 5z = 10$
   $-3x + 2z = 8$

To eliminate x from the 1st and 3rd equations, multiply the first equation by 3 and the third equation by 2 and add them together.

$$6x + 9y = 12$$
$$-6x + 4z = 16$$
$$9y + 4z = 28.$$

This gives a system of two equations in the two variables y and z:
$$4y - 5z = 10$$
$$9y + 4z = 28.$$

Multiply the 1st equation in this system by 4 and the second equation by 5, and add the resulting equations in order to eliminate z,

$$16y - 20z = 40$$
$$+45y + 20z = 140$$
$$61y = 180$$
$$y = \frac{180}{61} = 2.95082.$$

Substitute $y = 2.9508$ into equation 2 in the original system to get
$$4(2.95082) - 5z = 10$$
$$11.80328 - 5z = 10$$
$$-5z = -1.80328$$
$$z = 0.3607.$$

Now substitute $z = 0.3607$ into the 3rd equation to get
$$2(0.3607) - 3x = 8$$
$$0.7213 - 3x = 8$$
$$-3x = 7.2787$$
$$x = -2.4262.$$

The solution is $x = -2.4262, y = 2.9508, z = 0.3607$.

15. Solving the 3rd equation for z gives
$$z = \frac{5}{2} = 2.5.$$
Substitute 2.5 for z in the 2nd equation to get
$$0.002x - 0.088y + 0.04(2.5) = 0.0051$$
which simplifies to
$$0.002x - 0.088y = -0.0949.$$
This gives a system of two equations in two variables:
$$x + y = 5$$
$$0.002x - 0.088y = -0.0949.$$
Multiply the first equation by -2 and the second equation by 1000, getting
$$-2x - 2y = -10$$
$$2x - 88y = -94.9.$$
Add these equation to eliminate x
$$-90y = -104.9$$
$$y = 1.1656.$$
Now substitute 1.1656 for y in the first equation to get
$$x + 1.1656 = 5$$
$$x = 3.8344.$$
The solution is $x = 3.8344, y = 1.1656, z = 2.5.$

Section 7.2

1. $\begin{bmatrix} 3 & 1 \\ 2 & 6 \\ 1 & 3 \end{bmatrix} + \begin{bmatrix} 0 & 2 \\ 1 & -1 \\ 7 & 5 \end{bmatrix} = \begin{bmatrix} 3+0 & 1+2 \\ 2+1 & 6+(-1) \\ 1+7 & 3+5 \end{bmatrix} = \begin{bmatrix} 3 & 3 \\ 3 & 5 \\ 8 & 8 \end{bmatrix}$.

3. $\begin{bmatrix} 1 & 3 \\ 2 & 4 \end{bmatrix} - \begin{bmatrix} 4 & 3 \\ 2 & 1 \end{bmatrix} = \begin{bmatrix} 1-4 & 3-3 \\ 2-2 & 4-1 \end{bmatrix} = \begin{bmatrix} -3 & 0 \\ 0 & 3 \end{bmatrix}$.

5. $5 \begin{bmatrix} 2 & 1 \\ 3 & 6 \end{bmatrix} = \begin{bmatrix} 5(2) & 5(1) \\ 5(3) & 5(6) \end{bmatrix} = \begin{bmatrix} 10 & 5 \\ 15 & 30 \end{bmatrix}$.

7. $2\begin{bmatrix} 3 & 1 \\ 2 & 0 \end{bmatrix} - 4\begin{bmatrix} 0 & 2 \\ 1 & 3 \end{bmatrix} = \begin{bmatrix} 2(3) & 2(1) \\ 2(2) & 2(0) \end{bmatrix} - \begin{bmatrix} 4(0) & 4(2) \\ 4(1) & 4(3) \end{bmatrix} =$
$\begin{bmatrix} 6 & 2 \\ 4 & 0 \end{bmatrix} - \begin{bmatrix} 0 & 8 \\ 4 & 12 \end{bmatrix} = \begin{bmatrix} 6-0 & 2-8 \\ 4-4 & 0-12 \end{bmatrix} = \begin{bmatrix} 6 & -6 \\ 0 & -12 \end{bmatrix}$.

9. Not defined. The matrices are not the same size.

11. $\begin{bmatrix} 1 & 3 \\ -2 & 1 \end{bmatrix} + \begin{bmatrix} 1 & 2 \\ 3 & 4 \end{bmatrix} = \begin{bmatrix} 1+1 & 3+2 \\ -2+3 & 1+4 \end{bmatrix} = \begin{bmatrix} 2 & 5 \\ 1 & 5 \end{bmatrix}$.

13. $\begin{bmatrix} 1 & 2 \\ 3 & 1 \end{bmatrix}\begin{bmatrix} 5 \\ 2 \end{bmatrix} = \begin{bmatrix} 1(5)+2(2) \\ 3(5)+1(2) \end{bmatrix} = \begin{bmatrix} 9 \\ 17 \end{bmatrix}$.

15. Not defined. The first matrix has 3 columns and the second matrix has only 2 rows.

17. $\begin{bmatrix} 2 & 3 \\ 1 & 5 \end{bmatrix}\begin{bmatrix} 2 \\ 1 \end{bmatrix} = \begin{bmatrix} 2(2)+3(1) \\ 1(2)+5(1) \end{bmatrix} = \begin{bmatrix} 7 \\ 7 \end{bmatrix}$.

19. $\begin{bmatrix} 1 & 0 & 0 \\ 0 & 1 & 0 \\ 0 & 0 & 1 \end{bmatrix}\begin{bmatrix} 2 & 3 & 7 \\ 3 & 1 & 2 \\ 1 & 5 & 8 \end{bmatrix} = \begin{bmatrix} 2+0+0 & 3+0+0 & 7+0+0 \\ 0+3+0 & 0+1+0 & 0+2+0 \\ 0+0+1 & 0+0+5 & 0+0+8 \end{bmatrix} = \begin{bmatrix} 2 & 3 & 7 \\ 3 & 1 & 2 \\ 1 & 5 & 8 \end{bmatrix}$.

21. Not defined. It is not possible to multiply a 2 × 2 matrix times a 3 × 2 matrix.

23. Not defined. It is not possible to multiply a 3 × 2 matrix times a 3 × 3 matrix.

25. $\begin{bmatrix} 1 & 0 & -1 \\ 4 & 2 & -1 \\ 0 & 3 & 2 \end{bmatrix}\begin{bmatrix} 7 & 2 & 0 \\ -3 & 0 & 1 \\ 5 & -1 & 2 \end{bmatrix} = \begin{bmatrix} 7+(0)+(-5) & 2+0+1 & 0+0+(-2) \\ 28+(-6)+(-5) & 8+0+1 & 0+2+(-2) \\ 0+(-9)+10 & 0+0+(-2) & 0+3+4 \end{bmatrix} =$

$\begin{bmatrix} 2 & 3 & -2 \\ 17 & 9 & 0 \\ 1 & -2 & 7 \end{bmatrix}$

27. If the product of these two 2 × 2 matrices is the 2 × 2 identity matrix, then they are inverses of each other.

$\begin{bmatrix} 1 & 3 \\ 2 & 6 \end{bmatrix}\begin{bmatrix} 4 & -3 \\ -1 & 1 \end{bmatrix} = \begin{bmatrix} 4+(-3) & -3+3 \\ 8+(-6) & -6+6 \end{bmatrix} = \begin{bmatrix} 1 & 0 \\ 2 & 0 \end{bmatrix}.$

These two matrices are not inverses of each other.

29. If the product of these two 3 × 3 matrices is the 3 × 3 identity matrix, then they are inverses of each other.

$\begin{bmatrix} 1 & 2 & 3 \\ 2 & 5 & 3 \\ 1 & 0 & 8 \end{bmatrix}\begin{bmatrix} -40 & 16 & 9 \\ 13 & -5 & -3 \\ 5 & -2 & -1 \end{bmatrix} = \begin{bmatrix} -40+26+15 & 16+(-10)+(-6) & 9+(-6)+(-3) \\ -80+65+15 & 32+(-25)+(-6) & 18+(-15)+(-3) \\ -40+0+40 & 16+0+(-16) & 9+0+(-8) \end{bmatrix}$

$= \begin{bmatrix} 1 & 0 & 0 \\ 0 & 1 & 0 \\ 0 & 0 & 1 \end{bmatrix}.$

These two matrices are inverses of each other.

Section 7.3

1. The system of equations can be represented by the matrices:

   coefficient matrix $A = \begin{bmatrix} 1 & 1 \\ 3 & 2 \end{bmatrix}$; constant matrix $B = \begin{bmatrix} 2 \\ 5 \end{bmatrix}$;

   variable matrix $X = \begin{bmatrix} x \\ y \end{bmatrix}$.

   The solution matrix is determined by $X = A^{-1}B = \begin{bmatrix} 1 \\ 1 \end{bmatrix}$, so $x = 1$, and $y = 1$.

3. The system of equations can be represented by the matrices:

   coefficient matrix $A = \begin{bmatrix} 7 & 1 \\ 3 & 4 \end{bmatrix}$; constant matrix $B = \begin{bmatrix} 1 \\ -1 \end{bmatrix}$;

   variable matrix $X = \begin{bmatrix} x \\ y \end{bmatrix}$.

   The solution matrix is determined by $X = A^{-1}B = \begin{bmatrix} 0.2 \\ -0.4 \end{bmatrix}$, so $x = 0.2$ and $y = -0.4$.

5. The system of equations can be represented by the matrices:

   coefficient matrix $A = \begin{bmatrix} 3 & 2 \\ 1 & 4 \end{bmatrix}$; constant matrix $B = \begin{bmatrix} 8 \\ 5 \end{bmatrix}$;

   variable matrix $X = \begin{bmatrix} x \\ y \end{bmatrix}$.

   The solution matrix is determined by $X = A^{-1}B = \begin{bmatrix} 2.2 \\ 0.7 \end{bmatrix}$, so $x = 2.2$ and $y = 0.7$.

7. The system of equations can be represented by the matrices:

   coefficient matrix $A = \begin{bmatrix} 8 & 3 \\ -11 & 2 \end{bmatrix}$; constant matrix $B = \begin{bmatrix} 17 \\ 3 \end{bmatrix}$;

   variable matrix $X = \begin{bmatrix} x \\ y \end{bmatrix}$.

   The solution matrix is determined by $X = A^{-1}B = \begin{bmatrix} 0.5102 \\ 4.3061 \end{bmatrix}$, so $x = 0.5102$ and $y = 4.3061$.

9. The system of equations can be represented by the matrices:

coefficient matrix $A = \begin{bmatrix} 125 & -354 \\ 476 & 333 \end{bmatrix}$; constant matrix $B = \begin{bmatrix} 998 \\ 1026 \end{bmatrix}$;

variable matrix $X = \begin{bmatrix} x \\ y \end{bmatrix}$.

The solution matrix is determined by $X = A^{-1}B = \begin{bmatrix} 3.3101 \\ -1.6504 \end{bmatrix}$, so $x = 3.3101$ and $y = -1.6504$.

11. The system of equations can be represented by the matrices:

coefficient matrix $A = \begin{bmatrix} 1 & -3 & -2 \\ 2 & 8 & 9 \\ 1 & 5 & 6 \end{bmatrix}$; constant matrix $B = \begin{bmatrix} 11 \\ 23 \\ 13 \end{bmatrix}$;

variable matrix $X = \begin{bmatrix} x \\ y \\ z \end{bmatrix}$.

The solution matrix is determined by $X = A^{-1}B = \begin{bmatrix} 9.25 \\ -2.25 \\ 2.5 \end{bmatrix}$, so $x = 9.25, y = -2.25$, and $z = 2.5$.

13. The system of equations can be represented by the matrices:

coefficient matrix $A = \begin{bmatrix} 1 & 7 & 4 \\ -3 & -1 & 16 \\ 2 & 4 & -1 \end{bmatrix}$; constant matrix $B = \begin{bmatrix} -3 \\ 7 \\ 1 \end{bmatrix}$;

variable matrix $X = \begin{bmatrix} x \\ y \\ z \end{bmatrix}$.

The solution matrix is determined by $X = A^{-1}B = \begin{bmatrix} 4.66 \\ -1.78 \\ 1.2 \end{bmatrix}$, so $x = 4.66, y = -1.78$, and $z = 1.2$.

15. The system of equations can be represented by the matrices:

coefficient matrix $A = \begin{bmatrix} 5 & 2 & -1 \\ 4 & -3 & 2 \\ 3 & 7 & -5 \end{bmatrix}$; constant matrix $B = \begin{bmatrix} 4 \\ 5 \\ 8 \end{bmatrix}$;

variable matrix $X = \begin{bmatrix} x \\ y \\ z \end{bmatrix}$.

The solution matrix is determined by $X = A^{-1}B = \begin{bmatrix} 1.35 \\ -5.9 \\ -9.05 \end{bmatrix}$, so $x = 1.35, y = -5.9$, and $z = -9.05$.

17. The system of equations can be represented by the matrices:

coefficient matrix $A = \begin{bmatrix} 1 & 0 & 2 \\ 2 & 1 & 0 \\ 0 & 3 & 5 \end{bmatrix}$; constant matrix $B = \begin{bmatrix} -1 \\ 14 \\ -3 \end{bmatrix}$;

variable matrix $X = \begin{bmatrix} x \\ y \\ z \end{bmatrix}$.

The solution matrix is determined by $X = A^{-1}B = \begin{bmatrix} 5 \\ 4 \\ -3 \end{bmatrix}$, so $x = 5, y = 4$, and $z = -3$.

19. The system of equations can be represented by the matrices:

coefficient matrix $A = \begin{bmatrix} -\frac{1}{2} & -\frac{3}{4} & \frac{2}{3} \\ 2 & 0 & -5 \\ 1 & 1 & 1 \end{bmatrix}$; constant matrix $B = \begin{bmatrix} -1 \\ 6 \\ 3 \end{bmatrix}$;

variable matrix $X = \begin{bmatrix} x \\ y \\ z \end{bmatrix}$.

The solution matrix is determined by $X = A^{-1}B = \begin{bmatrix} 3.6122 \\ -0.85714 \\ 0.2449 \end{bmatrix}$, so $x = 3.6122, y = -0.85714$, and $z = 0.2449$.

Chapter 10 Exponential and Logarithmic Functions

Section 10.1

1. $10^{0.259} = 1.8155$

3. $e^2 = 7.3891$

5. $7.31(5.6)^{9.74} = 141,667,020.5$

7. $(2.66)^{-0.3} = 0.74565$

9. $-15.2(1.01)^{1.01} = -15.354$

11.

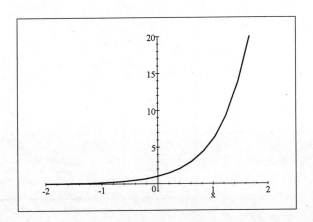

y-intercept: $(0, 1)$

13.

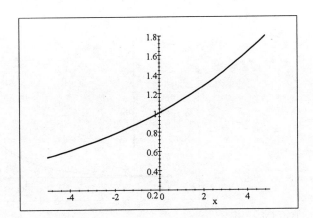

y-intercept: $(0, 1)$

15.

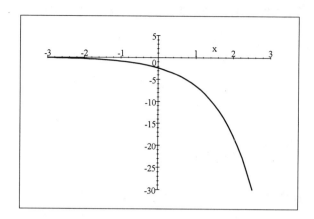

y-intercept: $(0, -2.4)$

17.

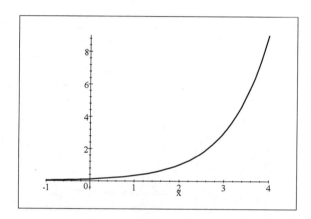

y-intercept: $\left(0, \frac{1}{9}\right)$

19.

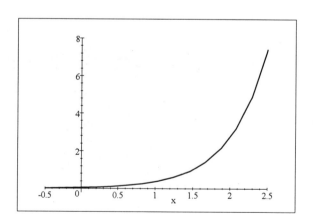

y-intercept: $(0, 0.049787)$

Section 10.2

1. Replace $f(x)$ by y

$$y = 2x - 1.$$

Solve for x:

$$x = \frac{y+1}{2};$$

hence $f^{-1}(y) = \frac{y+1}{2}$.

Checking the composites:

$$\begin{aligned} f^{-1}(f(x)) &= f^{-1}(2x-1) \\ &= \frac{(2x-1)+1}{2} \\ &= \frac{2x}{2} = x; \end{aligned}$$

also

$$\begin{aligned} f(f^{-1}(y)) &= f\left(\frac{y+1}{2}\right) \\ &= 2\left(\frac{y+1}{2}\right) - 1 \\ &= y + 1 - 1 = y. \end{aligned}$$

3. Replace $g(x)$ by y

$$y = \frac{x}{5} + 20.$$

Solve for x:

$$\begin{aligned} x &= 5(y - 20) \\ &= 5y - 100. \end{aligned}$$

Hence $g^{-1}(y) = 5y - 100$.

Checking the composites:

$$\begin{aligned} g^{-1}(g(x)) &= g^{-1}\left(\frac{x}{5} + 20\right) \\ &= 5\left(\frac{x}{5} + 20\right) - 100 \\ &= x + 100 - 100 = x; \end{aligned}$$

also

$$\begin{aligned} g(g^{-1}(y)) &= g(5y - 100) \\ &= \frac{5y - 100}{5} + 20 \\ &= y - 20 + 20 = y. \end{aligned}$$

5. Replace $f(x)$ by y

$$y = -5.4x$$

Solve for x:

$$x = -\frac{y}{5.4}.$$

Hence $f^{-1}(y) = -\frac{y}{5.4}$.
Checking the composites:

$$\begin{aligned} f^{-1}(f(x)) &= f^{-1}(-5.4x) \\ &= -\frac{-5.4x}{5.4} \\ &= x; \end{aligned}$$

also

$$\begin{aligned} f(f^{-1}(y)) &= f\left(-\frac{y}{5.4}\right) \\ &= -5.4\left(-\frac{y}{5.4}\right) \\ &= y. \end{aligned}$$

7. Replace $g(x)$ by y

$$y = -\sqrt{2}x + 3.$$

Solve for x:

$$x = \frac{y-3}{-\sqrt{2}} = -\frac{\sqrt{2}}{2}(y-3).$$

Hence $g^{-1}(y) = -\frac{\sqrt{2}}{2}(y-3)$.
Checking the composites:

$$\begin{aligned} g^{-1}(g(x)) &= g^{-1}\left(-\sqrt{2}x + 3\right) \\ &= -\frac{\sqrt{2}}{2}\left(\left(-\sqrt{2}x+3\right) - 3\right) \\ &= -\frac{\sqrt{2}}{2}\left(-\sqrt{2}x\right) = x; \end{aligned}$$

also

$$\begin{aligned} g(g^{-1}(y)) &= g\left(-\frac{\sqrt{2}}{2}(y-3)\right) \\ &= -\sqrt{2}\left(-\frac{\sqrt{2}}{2}(y-3)\right) + 3 \\ &= y - 3 + 3 = y. \end{aligned}$$

9. Replace $i(x)$ by y

$$y = 1 - 2.37x.$$

Solve for x:

$$x = \frac{y-1}{-2.37}$$
$$= -0.42194y + 0.42194$$

hence $i^{-1}(y) = -0.42194y + 0.42194$.
Checking the composites:

$$i^{-1}(i(x)) = i^{-1}(1 - 2.37x)$$
$$= -0.42194(1 - 2.37x) + 0.42194$$
$$= -0.42194 + x + 0.42194 = x.$$

also

$$i(i^{-1}(y)) = i(-0.42194y + 0.42194)$$
$$= 1 - 2.37(-0.42194y + 0.42194)$$
$$= 1 + y - 1 = y.$$

11. Replace $k(x)$ by y

$$y = 3x^3 + 2.$$

Solve for x:

$$x^3 = \frac{y-2}{3}$$
$$x = \sqrt[3]{\frac{y-2}{3}};$$

hence $k^{-1}(y) = \sqrt[3]{\frac{y-2}{3}}$.
Checking the composites:

$$k^{-1}(k(x)) = k^{-1}(3x^3 + 2)$$
$$= \sqrt[3]{\frac{(3x^3 + 2) - 2}{3}}$$
$$= \sqrt[3]{\frac{3x^3}{3}} = \sqrt[3]{x^3} = x;$$

also

$$k(k^{-1}(y)) = k\left(\sqrt[3]{\frac{y-2}{3}}\right)$$
$$= 3\left(\sqrt[3]{\frac{y-2}{3}}\right)^3 + 2$$
$$= 3\left(\frac{y-2}{3}\right) + 2$$
$$y - 2 + 2 = y.$$

13. Replace $f(x)$ by y and solve for x

$$y = 1 - x^2$$
$$x^2 = 1 - y$$
$$x = \pm\sqrt{1-y}.$$

$x = \pm\sqrt{1-y}$ is not a function, since there are two values of x for one y.

Section 10.3

1. $\log_3 81 = 4$ because $3^4 = 81$.

3. $\log_{10} 10,000 = 4$ because $10^4 = 10,000$.

5. $\log_7 1 = 0$ because $7^0 = 1$.

7. $\log 17 = 1.2304$

9. $\ln 2 = 0.6931$

11. $\log 0.00028 = -3.5528$

13. $\ln \frac{1}{2} = -0.69315$

15. $\ln \frac{1}{e^4} = -4$

17. $y = 5 - 2\log x$

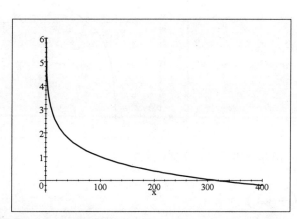

The x-intercept is approximately $(316.23, 0)$.

19. $y = 1 - 3.2 \ln x$

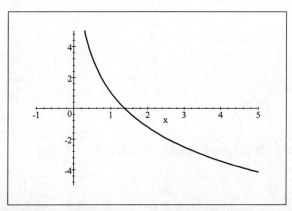

The x-intercept is approximately $(1.3668, 0)$.

21. $y = 3\log_{10} x + 1$

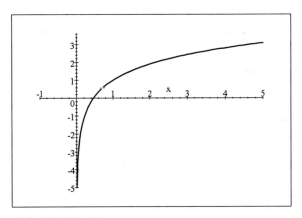

The x-intercept is approximately $(0.464, 0)$.

23. $y = 0.56 + 1.3\ln(x - 1)$

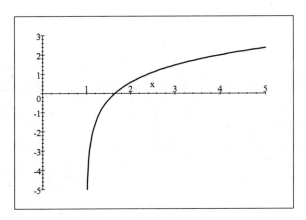

The x-intercept is approximately $(1.65, 0)$.

25. $y = 2.4 + \ln\left(\frac{x}{5} + 1\right)$

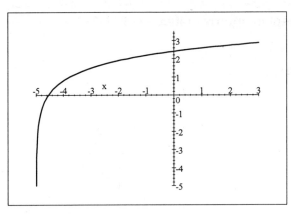

The x-intercept is approximately $(-4.546, 0)$.

Section 10.4

1. $f(t) = 1.2 + 3.7\ln t$

a.

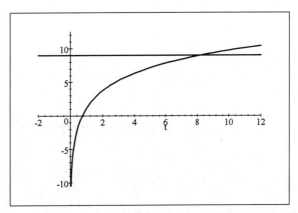

t is approximately 8.23 when $f(t) = 9$.

b.

$$1.2 + 3.7\ln t = 9$$
$$3.7\ln t = 7.8$$
$$\ln t = 2.10811$$
$$t = e^{2.10811} = 8.2327.$$

3. $H(s) = 0.002 - 0.035\ln s$

a.

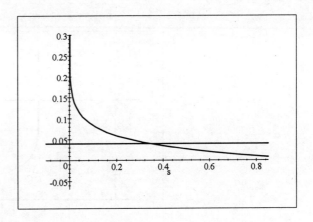

s is approximately 0.34 when $H(s) = 0.04$.

b.

$$0.002 - 0.035\ln s = 0.04$$
$$-0.035\ln s = 0.038$$
$$\ln s = -1.0857$$
$$s = e^{-1.0857} = 0.3377.$$

5. $r(t) = e^{8t-4}$
   a.

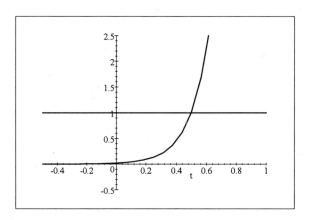

t is approximately 0.5 when $r(t) = 1$.

   b.
$$e^{8t-4} = 1$$
$$\ln(e^{8t-4}) = \ln(1)$$
$$8t - 4 = 0$$
$$8t = 4$$
$$t = 0.5.$$

7. $g(t) = 1.2 - \log_{10}(4t - 3)$
   a.

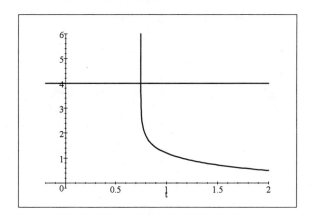

t is approximately equal to 0.75 when $g(t) = 4$.

b.

$$1.2 - \log(4t - 3) = 4$$
$$-\log(4t - 3) = 2.8$$
$$\log(4t - 3) = -2.8$$
$$10^{4t-3} = 10^{-2.8}$$
$$4t - 3 = 0.00158$$
$$4t = 3.00158$$
$$t = 0.7504$$

9. $y = 0.43 - 0.7\ln(x - 1)$

a.

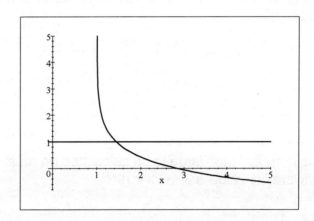

x is approximately 1.4 when $y = 1$.

b.

$$0.43 - 0.7\ln(x - 1) = 1$$
$$-0.7\ln(x - 1) = 0.57$$
$$\ln(x - 1) = -0.8143$$
$$x - 1 = e^{-0.8143}$$
$$x - 1 = 0.443$$
$$x = 1.443$$

11. $h(x) = \ln(3x)$

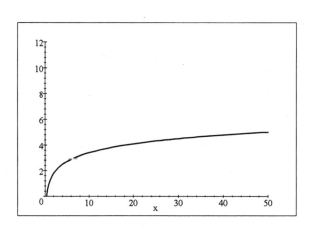

a. The graph shows that x is a very large number when $h(x) = 999$ and can not be found graphically.

b.

$$\ln(3x) = 999$$
$$3x = e^{999}$$
$$x = \frac{1}{3}e^{999}.$$

This number is too large to be calculated on most calculators. However, using computer software, its value is 2.4158×10^{433}.

13. $k(w) = 4 - 3^{w+1}$

a.

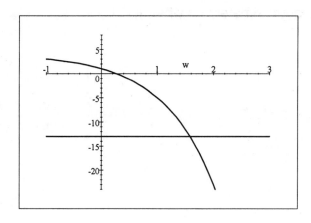

w is approximately equal to 1.58 when $k(w) = -13$.

62

b.

$$4 - 3^{w+1} = -13$$
$$-3^{w+1} = -17$$
$$3^{w+1} = 17$$
$$\ln(3^{w+1}) = \ln(17)$$
$$(w+1)\ln 3 = \ln 17$$
$$w + 1 = \frac{\ln 17}{\ln 3}$$
$$w = \frac{\ln 17}{\ln 3} - 1 = 1.5789.$$

15. $3 - 2^x = 7$

$-2^x = 4$

$2^x = -4$; no solution since $2^x > 0$ for all x.

Section 10.5

1. Replace $g(x)$ with y and solve for x:

$$y = 3^x$$
$$\ln y = \ln(3^x)$$
$$\ln y = x \ln 3$$
$$x = \frac{\ln y}{\ln 3} = 0.91024 \ln y.$$

Thus, $g^{-1}(y) = 0.91024 \ln y$;
$g(2) = 3^2 = 9; g^{-1}(9) = 0.91024 \ln 9 = 2.$

3. Replace $f(x)$ with y and solve for x:

$$y = 6.03^x$$
$$\ln y = \ln(6.03^x)$$
$$\ln y = x \ln 6.03$$
$$x = \frac{\ln y}{\ln 6.03} = 0.55656 \ln y.$$

Thus $f^{-1}(y) = 0.55656 \ln y$;
$f(2.5) = 6.03^{2.5} = 89.288; f^{-1}(89.288) = 0.55656 \ln 89.288 = 2.5.$

5. Replace $F(x)$ with y and solve for x:

$$y = 1 + 5^x$$
$$y - 1 = 5^x$$
$$\ln(y-1) = \ln 5^x$$
$$\ln(y-1) = x \ln 5$$
$$x = \frac{\ln(y-1)}{\ln(5)} = 0.62133 \ln(y-1).$$

Thus, $F^{-1}(y) = 0.62133 \ln(y-1)$;
$F(0) = 1 + 5^0 = 2; F^{-1}(2) = 0.62133 \ln(2-1) = 0.$

7. Replace $g(x)$ with y and solve for x:

$$y = \frac{1.01^x}{5}$$
$$5y = 1.01^x$$
$$\ln(5y) = \ln(1.01^x)$$
$$\ln(5y) = x \ln(1.01)$$
$$x = \frac{\ln(5y)}{\ln(1.01)} = 100.5 \ln(5y).$$

Thus, $g^{-1}(y) = 100.5 \ln(5y)$;
$g(7) = 0.214427; g^{-1}(0.214427) = 7.$

9. Replace $B(x)$ with y and solve for x:

$$y = -15(3.5)^{x+1}$$
$$-\frac{y}{15} = 3.5^{x+1}$$
$$\ln\left(-\frac{y}{15}\right) = \ln(3.5^{x+1})$$
$$\ln\left(-\frac{y}{15}\right) = (x+1)\ln 3.5$$
$$\frac{\ln\left(-\frac{y}{15}\right)}{\ln 3.5} = x+1$$
$$x = \frac{\ln\left(-\frac{y}{15}\right)}{\ln 3.5} - 1$$
$$x = \frac{\ln\left(-\frac{y}{15}\right) - \ln 3.5}{\ln 3.5}$$
$$x = 0.79824\ln(-0.019048y)$$

Thus, $B^{-1}(y) = 0.79824\ln(-0.019048y)$;
$B(4) = -7878.28125; B^{-1}(-7878.28125) = 4$.

11. Replace $f(x)$ with y and solve for x:

$$y = 2.7 + 1.3\ln x$$
$$y - 2.7 = 1.3\ln x$$
$$\frac{y-2.7}{1.3} = \ln x$$
$$x = e^{\frac{y-2.7}{1.3}}$$
$$x = e^{0.76923y - 2.0769}$$

Thus, $f^{-1}(y) = e^{0.76923y-2.0769}$; $f(2) = 3.601; f^{-1}(3.601) = 2$.

13. Replace $h(x)$ with y and solve for x:

$$y = 1.5\ln(3x) - 9$$
$$y + 9 = 1.5\ln(3x)$$
$$\frac{y+9}{1.5} = \ln(3x)$$
$$e^{\frac{y+9}{1.5}} = e^{\ln(3x)}$$
$$e^{\frac{y+9}{1.5}} = 3x$$
$$x = \frac{1}{3}e^{\frac{y+9}{1.5}} = 0.33333e^{0.66667y+6.0}$$

Thus, $h^{-1}(y) = 0.33333e^{0.66667y+6.0}$.
Checking the composites:

$$h(h^{-1}(y)) = h\left(\frac{1}{3}e^{\frac{y+9}{1.5}}\right) = 1.5\ln\left(3\left(\frac{1}{3}e^{\frac{y+9}{1.5}}\right)\right) - 9 = 1.5\ln\left(e^{\frac{y+9}{1.5}}\right) - 9$$
$$= 1.5\left(\frac{y+9}{1.5}\right) - 9 = y + 9 - 9 = y.$$

Also,
$$h^{-1}(h(x)) = h^{-1}(1.5\ln(3x) - 9) = \frac{1}{3}e^{\frac{(1.5\ln(3x)-9)+9}{1.5}} = \frac{1}{3}e^{\ln(3x)} = \frac{1}{3}(3x) = x.$$

$h(8) = -4.2329; h^{-1}(-4.2329) = 8.$

15. Replace $U(x)$ with y and solve for x:

$$y = 0.874\ln(2x)$$
$$\frac{y}{0.874} = \ln(2x)$$
$$2x = e^{\frac{y}{0.874}}$$
$$x = \frac{1}{2}e^{\frac{y}{0.874}} = 0.5e^{1.1442y}.$$

Thus $U^{-1}(y) = 0.5e^{1.1442y}$

Checking the composites:
$$U(U^{-1}(y)) = U(0.5e^{1.1442y}) = 0.874\ln\left(\frac{1}{2}(2e^{1.1442y})\right) = 0.874\ln(e^{1.1442y})$$
$$= 0.874(1.1442y) = y.$$

$U(-3)$ is undefined since the domain of $y = \ln x$ is $x \geq 0$.

17. Replace $H(x)$ with y and solve for x:

$$y = 2.3 - \log_{2.3} x$$
$$\log_{2.3} x = 2.3 - y$$
$$x = 2.3^{2.3-y}.$$

Thus, $H^{-1}(y) = 2.3^{2.3-y}$.

Checking the composites:
$$H(H^{-1}(y)) = H(2.3^{2.3-y}) = 2.3 - \log_{2.3}(2.3^{2.3-y}) = 2.3 - (2.3 - y) = y;$$

Also,
$$H^{-1}(H(x)) = H^{-1}(2.3 - \log_{2.3} x) = 2.3^{2.3-(2.3-\log_{2.3} x)} = 2.3^{\log_{2.3} x} = x.$$

$H(0.1) = 2.3 - \log_{2.3}(0.1) = 5.0645; H^{-1}(5.0645) = 0.1.$

19. Replace $K(x)$ with y and solve for x:

$$y = \log(2x+5) + 4$$
$$y - 4 = \log(2x+5) = \frac{\ln(2x+5)}{\ln 10}$$
$$2.3026(y-4) = \ln(2x+5)$$
$$e^{2.3026(y-4)} = e^{\ln(2x+5)}$$
$$0.0001 e^{2.3026y} = 2x + 5$$
$$x = 0.0005 e^{2.3026y} - 2.5.^*$$

Thus, $K^{-1}(y) = 0.0005 e^{2.3026y} - 2.5$.

Checking the composites:

$$K(K^{-1}(y)) = K\left(\frac{10^{(y-4)} - 5}{2}\right) = \log\left(2\left(\frac{10^{(y-4)} - 5}{2}\right) + 5\right) + 4 = \log(10^{(y-4)}) + 4 = (y-4) + 4 = y.$$

Also

$$K^{-1}(K(x)) = K^{-1}(\log(2x+5) + 4) = \frac{10^{(\log(2x+5) + 4 - 4)} - 5}{2} = \frac{10^{\log(2x+5)} - 5}{2} = \frac{(2x+5) - 5}{2} = x.$$

$K(-2) = 4; K^{-1}(4) = -2.$

*Alternate solution:

$$y = \log(2x+5) + 4$$
$$y - 4 = \log(2x+5)$$
$$10^{y-4} = 2x + 5$$
$$10^{y-4} - 5 = 2x$$
$$x = \frac{1}{2}(10^y 10^{-4} - 5)$$
$$x = 0.0005(10^y) - 2.5.$$

Chapter 12 Geometric Series

Section 12.1

1. The geometric series
$$1 + 6 + 36 + 216 + 1296 + 7776 + 46,656 + 279,936$$
can be rewritten as
$$1(6^0) + 1(6^1) + 1(6^2) + 1(6^3) + 1(6^4) + 1(6^5) + 1(6^6) + 1(6^7).$$
Hence $a = 1, r = 6, n = 7$.
The sum of the series is
$$S = \frac{1(6^{7+1} - 1)}{6 - 1} = 335,923.$$

3. $a = 71.2, r = 4.3, n = 12$;
The sum of the series is
$$S = \frac{71.2(4.3^{12+1} - 1)}{4.3 - 1} = 3,707,284,998.$$

5. $a = 0.5, r = 0.1, n = 793$;
$$S = \frac{0.5(0.1^{793+1} - 1)}{0.1 - 1} = 0.5556.$$

7. $a = 1.5, r = e, n = 35$;
$$S = \frac{1.5(e^{35+1} - 1)}{e - 1} = 3.7636 \times 10^{15}.$$

9. The geometric series is
$$3 + 3(8)^1 + 3(8)^2 + 3(8)^3 + 3(8)^4 + 3(8)^5.$$
The sum is
$$S = \frac{3(8^{5+1} - 1)}{8 - 1} = 112,347.$$

11. The geometric series is
$$2.16 + 2.16(1.7)^1 + 2.16(1.7)^2 + \ldots + 2.16(1.7)^{52}.$$
The sum is
$$S = \frac{2.16(1.7^{52+1} - 1)}{1.7 - 1} = 5.0483 \times 10^{12}.$$

13. The geometric series is
$$2 + 2(3)^1.$$
The sum is
$$S = \frac{2(3^{1+1} - 1)}{3 - 1} = 8.$$

15. The geometric series is
$$0.15 - 0.15\left(\tfrac{3}{4}\right) + 0.15\left(\tfrac{3}{4}\right)^2 - 0.15\left(\tfrac{3}{4}\right)^3 + 0.15\left(\tfrac{3}{4}\right)^4 - 0.15\left(\tfrac{3}{4}\right)^5 + 0.15\left(\tfrac{3}{4}\right)^6 - 0.15\left(\tfrac{3}{4}\right)^7.$$
The sum is
$$S = \frac{0.15\left(\left(-\tfrac{3}{4}\right)^{7+1} - 1\right)}{-\tfrac{3}{4} - 1} = 0.077133.$$

Chapter 17 Linear Inequalities in Two Variables and Systems of Inequalities

(The graphs in this chapter were created on the TI-83 calculator with the indicated viewing window.)

Section 17.1

1. First solve the inequality for y in terms of x:

$$5x - 2y \geq 12$$
$$2y \geq 12 - 5x$$
$$y \geq 6 - \frac{5}{2}x$$

   Next graph the line $y = 6 - \frac{5}{2}x$.
   The set of points (x,y) that lie on this line is the set of all (x,y) such that y is exactly equal to $6 - \frac{5}{2}x$. These points make up part of the set of solutions to the linear inequality but not all. Also y can be greater than $6 - \frac{5}{2}x$, so all points above that line would also be solutions.

   Window $[-5, 5, 1], [-5, 10, 1]$

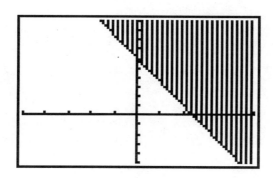

3. First solve the inequality for y in terms of x:

$$0.1x - 2.5y \geq 1.4$$
$$-2.5y \geq 1.4 - 0.1x$$
$$y \leq \frac{1.4}{-2.5} - \frac{0.1}{-2.5}x$$
$$y \leq -0.56 + 0.04x.$$

   Next, graph the line $y = -0.56 + 0.04x$.
   The set of points (x,y) which lie on this line is the set of all (x,y) such that y is exactly equal to $0.04x - 0.56$. These points make up part of the solution to the inequality but not all. Also, y can also be less than $0.04x - 0.56$, so all points below that line would be solutions.

Window $[-5, 25, 1], [-2, 2, 1]$

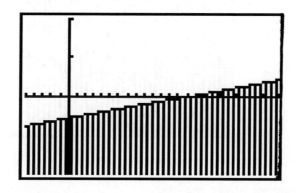

5. First solve the inequality for y in terms of x:

$$10.2x + 4.5y \leq 14.1$$
$$4.5y \leq -10.2x + 14.1$$
$$y \leq -2.267x + 3.133.$$

Next, graph the line $y = -2.267x + 3.133$.

The set of points (x,y) which lie on this line is the set of all (x,y) such that y is exactly equal to $-2.267 + 3.133$. These points make up part of the set of solutions to the inequality but not all. Also, y can be less than $-2.267x + 3.133$, so all points below that line would also be solutions.

Window $[-5, 5, 1], [-5, 5, 1]$

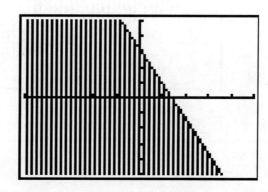

7. Graph the vertical line $x = 3$. The points on the line are the set of points (x,y) such that $x = 3$. In this case, they are not part of the solution set. Only the points lying to the left of the line satisfy the inequality $x < 3$.

Window $[-5, 5, 1], [-5, 5, 1]$

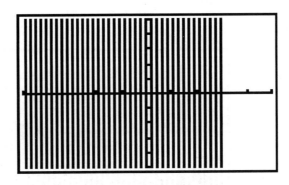

9. First solve the inequality for y in terms of x.
$$3x - 7 \leq 2y + 4x$$
$$-x - 7 \leq 2y$$
$$y \geq -\frac{1}{2}x - \frac{7}{2}.$$

Next, graph the line $y = -\frac{1}{2}x - \frac{7}{2}$.

The set of points (x,y) which lie on this line is the set of all (x,y) such that y is exactly equal to $-\frac{1}{2}x - \frac{7}{2}$. These points make up part of the set of solutions to the inequality but not all. Also, y can be greater than $-\frac{1}{2}x - \frac{7}{2}$, so all points above that line would also be solutions.

Window $[-12, 10, 1], [-10, 10, 1]$

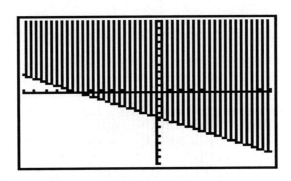

Section 17.2

The following graphs were created on the TI-83 calculator with the indicated viewing window.

1. The solution to the first inequality lies in the region on or above the line $x + y \geq 10$; the solution to the second inequality lies in the region on or above the line $2x - y \geq 2$. The solution set is the region common to both and is shown below. The corner is $(4, 6)$, the solution to the linear system

$$x + y = 10$$
$$2x - y = 2.$$

Window $[-5, 15, 1], [-5, 15, 1]$

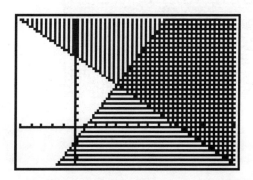

3. Graph each line and shade the appropriate region. The solution set is the shaded region shown below. The corner is $(7.33, -2.67)$, the solution (rounded) to the linear system

$$x - y = 10$$
$$2x + y = 12.$$

Window $[-5, 15, 1], [-10, 10, 1]$

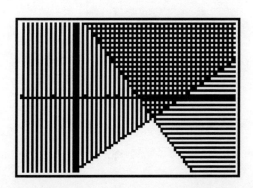

73

5. The solution to the first inequality is the region on or below the line $y = x$, the solution to the second inequality is the region on or above the line $y = \frac{1}{2}x - 1$, and the solution to the third inequality is the region on or below the line $y = 2 - x$. The solution set to the entire system is shown below with corners $(-2, -2), (2, 0),$ and $(1, 1)$.

Window $[-3, 3, 1], [-3, 3, 1]$

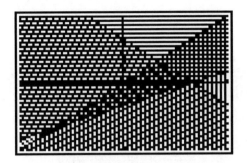

Section 17.3

1. The feasible region is shown below:

   Window $[0, 5, 1], [0, 5, 1]$

   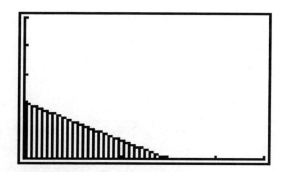

   Evaluate the objective function $E = x + y$ at the corners $(0, 2), (0, 0), (3, 0)$.

Corners	Value of E
(0, 0)	0 + 0
(0, 2)	0 + 2
(3, 0)	3 + 0

   The maximum value of the objective function is $E = 3$ which occurs at the point $(3, 0)$.

3. The graph of the feasible region is shown below.

   Window $[-10, 10, 1], [-10, 10, 1]$

   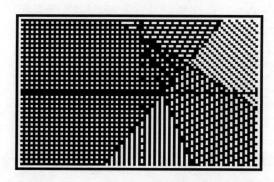

   Evaluate the objective function $G = 3x + 2y$ at the corners $(0, 8), (2, 0), (4, 4)$.

Corners	Value of G
(0, 8)	$3(0) + 2(8) = 16$
(2, 0)	$3(2) + 2(0) = 6$
(4, 4)	$3(4) + 2(4) = 20$.

The minimum value of the objective function is $G = 6$ which occurs at the point $(2, 0)$.

5. The graph of the feasible region is shown below.
   Window $[-10, 10, 1], [-10, 10, 1]$

Evaluate the objective function $J = 2x - y$ at the corners $(0, 5), (5, 0), (7, 0), (7, 12)$.

Corner	Value of J
(0, 5)	$J = 2(0) - 5 = -5$
(5, 0)	$J = 2(5) - 0 = 10$
(7, 0)	$J = 2(7) - 0 = 14$
(7, 12)	$J = 2(7) - 12 = 2$

The minimum value of the objective function is $J = -5$ which occurs at the point $(0, 5)$.